THE BIOGRAPHY OF NUMBERS:
ZERO

THE BIOGRAPHY OF NUMBERS:
ZERO

Kevin Cunningham

GREENSBORO, NORTH CAROLINA

 To join the discussion about this title, please check out the Morgan Reynolds Readers Club on Facebook, or Like our company page to stay up to date on the latest Morgan Reynolds news!

THE BIOGRAPHY OF NUMBERS:
ZERO

Copyright © 2014 by Morgan Reynolds Publishing

All rights reserved
This book, or parts therof, may not be reproduced
in any form except by written consent of the publisher.
For more information write:
Morgan Reynolds Publishing, Inc.
620 South Elm Street, Suite 387
Greensboro, NC 27406 USA

Library of Congress Cataloging-in-Publication Data

Cunningham, Kevin, 1966-
  Zero / by Kevin Cunningham. -- First edition.
      pages cm. -- (The biography of numbers)
  Includes bibliographical references and index.
   ISBN 978-1-59935-392-0 -- ISBN 978-1-59935-393-7 (e-book)
 1. Zero (The number) 2. Number concept. I. Title.
  QA241.C86 2013
  513.5--dc23
                                                2013005809

Printed in the United States of America
First Edition

Book cover and interior designed by:
Ed Morgan, navyblue design studio
Greensboro, NC

Contents

1 **Before Zero** 7
2 **A Universe in Crystal** 17
3 **Long Counts and Indian Numbers** 27
4 **The Messenger** 39
5 **The Zero Revolutions** 49

    Timeline 56
    Biographical Sketches 58
    Glossary 60
    Bibliography 61
    Web sites 62
    Index 63

CHAPTER 1

BEFORE ZERO

The Biography of Numbers

Anthropologists theorize that prehistoric peoples counted in only the simplest way. Studies of so-called "Stone Age" peoples that survived into the twentieth century suggest our ancestors probably conceived of two numerical concepts: one and many. Some of those ancestors may have counted as high as three before many took over.

Counting systems based in the number ten—perhaps inspired by the number of fingers on both human hands—only came along later to make the number many a bit more specific.

8

ZERO

If prehistoric peoples got by without seven or forty-one, they certainly did not need zero. Their worldview, as far as we know, concerned itself with material things—the sheep or spears or wives (or husbands) an individual possessed or at least knew enough about to want.

But what he or she lacked, though possibly worrisome to the person at the time, never inspired anyone to invent a number to represent a nothing. It would be thousands of years until humanity made that leap.

The Biography of Numbers

Babylon and on

Once humans began to build complex societies, however, more numbers became necessary. The idea of property required numbers to express distance to draw lines between, say, one farm and the next. Trade demanded number systems that allowed merchants and customers to track profit and loss, and weigh and measure volume. Numbers helped the ancient Egyptians build their pyramids and the ancient Chinese their long defensive walls.

Of the ancient cultures, the Egyptians may have been the first to flirt with the idea of zero. An accounting document from over 3,500 years ago shows a **hieroglyph** called *nfr* as the balance after subtracting payments from income. Clearly merchants valued reaching *nfr* in their business. *Nfr* in the ancient Egyptian language means "beautiful" or "complete."

The Pyramids at Giza

ZERO

The idea of *nfr* also served as the ground level starting point for buildings.

To the east, the ancestors of the people we call the Sumerians created the Sumer culture in Mesopotamia. The Sumerians listed property and kept inventories using a script called cuneiform. In use by 3,400 BCE, cuneiform is the world's first known writing system. Sumerians used a wedge-shaped tool to press the cuneiform symbols into moist clay tablets. The tablets, once dry, proved durable. Archaeologists study the thousands of tablets that survived until today.

Cuneiform writing on a stone tablet

11

The Biography of Numbers

Empty space

The city of Babylon, located in the same region as Sumer, dates to the 1800s BCE. About 1,200 years after its founding, Babylon became the independent hub of a larger Babylonian society and culture.

The Babylonians used two symbols, ︎❘ and ︎◁, in different combinations to represent numbers. Their intent was to represent numbers calculated on an **abacus**. This computing device allowed a user to calculate by sliding beads or rocks on wires held within a frame. Each wire represented a column in a number. An educated abacus reader could see the position of the beads or rocks and immediately understand the number being expressed.

An abacus

12

ZERO

Babylon, reconstructed in modern day Iraq

To make numbers clear, **scribes** writing cuneiform on tablets used spaces to keep numbers in specific columns. Human handwriting, alas, could be as sloppy in Babylon as in other times and places. Archaeologists have found that some scribes had a hard time writing spaces in a consistent way. In those cases, a tablet reader had to know enough about the business or a situation to make sense of the uncertain figures.

The Babylonian system invented a third symbol to clarify things.

The Biography of Numbers

Double wedge

Before or during the 500s BCE (though perhaps later), the Babylonians began to use the symbol ⪤ to indicate an empty space. This symbol of a double wedge kept the numerical symbols in hard-to-mistake columns that indicated **place value**.

Historians sometimes call ⪤ a zero. But the Babylonians, and the other peoples who adopted placeholder symbols (if they did—experts differ), had no conception of their zero as a number, in the way that we do. The ⪤ merely represented a space that aided reading comprehension. A person did not use the double wedge to create bigger numbers or to add or subtract or calculate in other ways.

Other methods

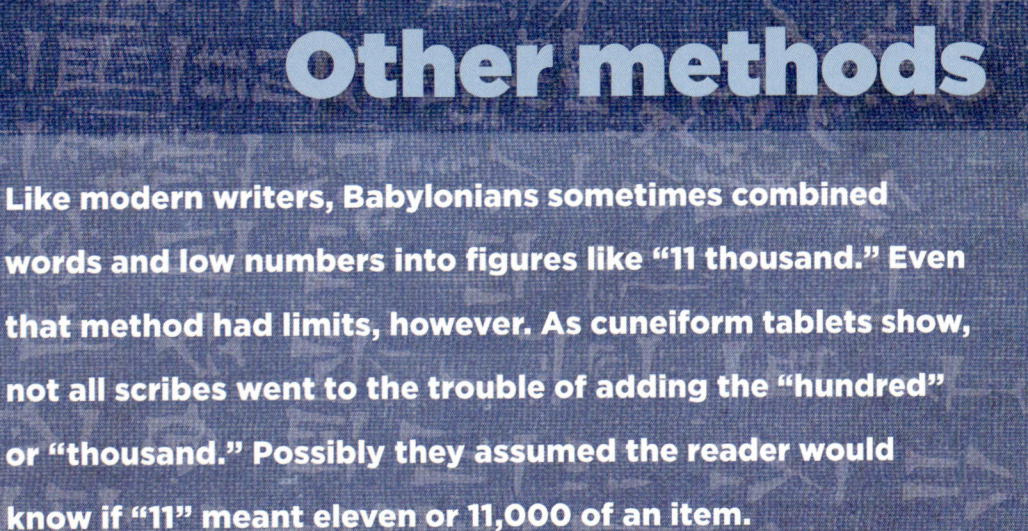

Like modern writers, Babylonians sometimes combined words and low numbers into figures like "11 thousand." Even that method had limits, however. As cuneiform tablets show, not all scribes went to the trouble of adding the "hundred" or "thousand." Possibly they assumed the reader would know if "11" meant eleven or 11,000 of an item.

CHAPTER 2

A UNIVERSE in CRYSTAL

The Babylonians' placeholder idea may have made its way to Greece via astronomy. What evidence exists of a Greek placeholder occurs in astronomical charts. Greek writers drew their placeholder as a version of omicron, the fifteenth letter in their alphabet. Omicron resembles our modern "o." Historians, however, think the similarity is a coincidence.

Greek astronomers did not use the placeholder much, if at all. In fact, the Greeks seemed to utterly reject the very idea of zero.

In 2000, science writer Charles Seife proposed that the Greeks wished away the idea of zero because it threatened their core beliefs about the universe. Zero suggested immense numbers—a never ending number line going on into **infinity**. At the same time zero symbolized a void, an absence of all matter, a nothing.

Seife believed that neither infinity nor the void appealed to Greek thinkers. Their view of the universe did not permit the ideas of infinite space and the absolute emptiness of a **vacuum**. Their view, in other words, did not permit the nothingness symbolized by zero.

The Parthenon in Athens, Greece

ZERO

The Biography of Numbers

The spheres

The philosopher Aristotle espoused the basics of the Greek belief in the heavens as a series of crystal spheres. Claudius Ptolemy, an astronomer and geographer who wrote in Greek and lived in Alexandria, Egypt, further shaped the ideas. In his *Almagest*, one of the most influential texts of ancient times, Ptolemy merged his own scientific work with knowledge culled from Aristotle and other ancient sources.

A bust of Aristotle

Claudius Ptolemy

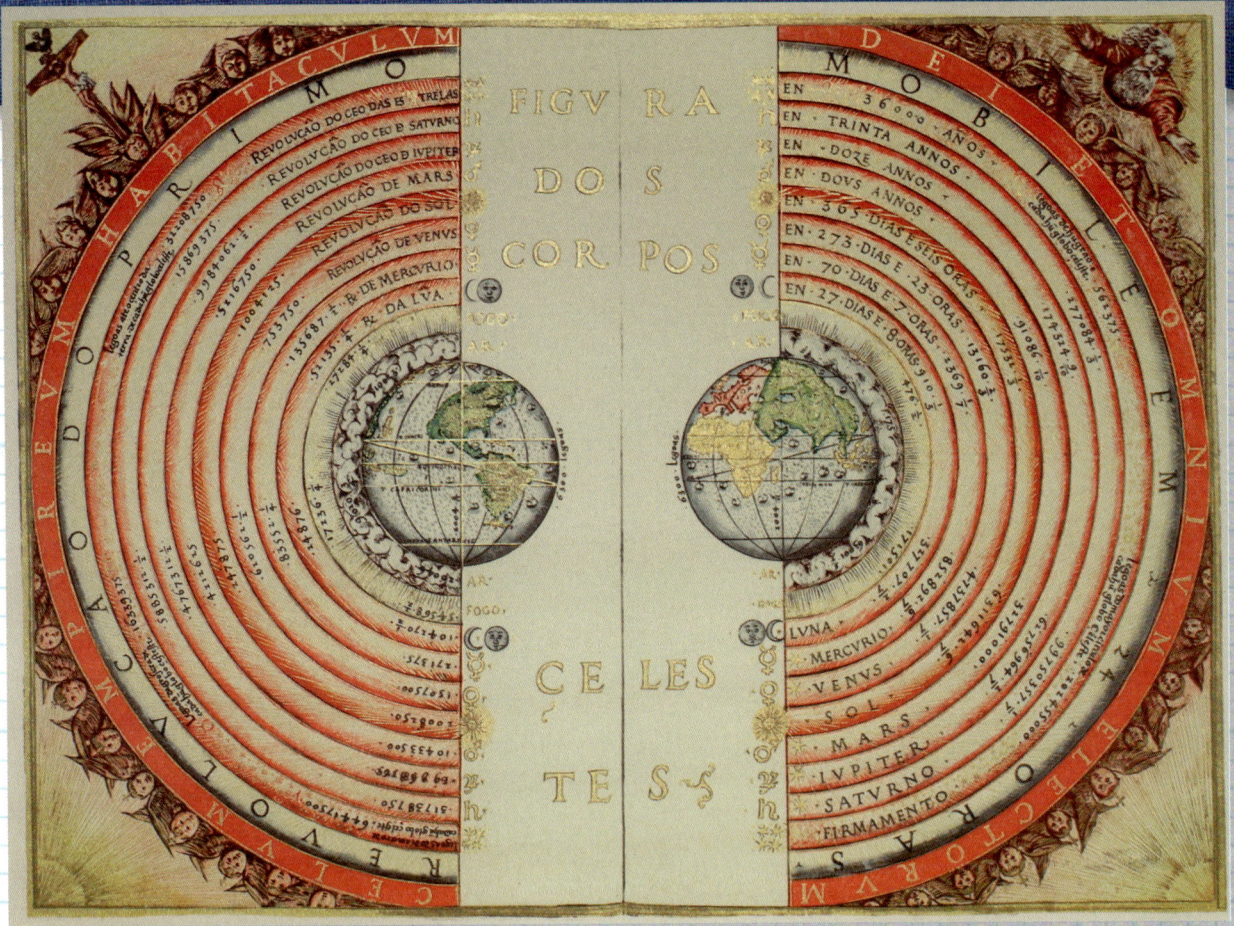

An illustration of the Ptolemaic geocentric system by Portuguese cosmographer and cartographer Bartolomeu Velho

Ptolemaic theory placed Earth, an unmoving sphere, at the center of a series of concentric spheres. The system of spheres made up the entire universe.

Ptolemy believed a heavenly body moved around two circles simultaneously. Take the Moon, the closest body to Earth in Ptolemy's system. The Moon rotated on a circular route of its own called the **epicycle**. The epicycle in turn moved around a larger crystal sphere. Ptolemy stated that Mercury traveled in the sphere beyond the Moon. Venus took the next sphere, then the Sun, then the rest of the planets known in Ptolemy's time. A final sphere housed the stars.

The Biography of Numbers

Prime movers

The system did not fully explain the movements astronomers could see with their own eyes. Ptolemy, a genuine scientist, adjusted his ideas when he recognized his theory failed to explain his own observations. Both he and scientists who followed him added refinements to make the system fit their facts.

The Aristotle-Ptolemy system had far reaching effects on faith. Aristotle proposed, and Western thinkers of ancient times agreed, that the crystal spheres moved. The second sphere with Mercury moved the first that held the Moon, the theory stated. Venus's sphere then moved Mercury's, and out and out.

Full Moon
Mars
Venus

ZERO

Infrared photo of the galaxy

Ptolemy and his followers, influenced by Greek culture and in some cases Greek themselves, were hostile to the idea of infinity. The last crystal sphere holding the stars was, and had to be, the end of the universe. No infinite space filled by a vacuum existed or could exist beyond it.

But what, then, put the universe in motion? Aristotle proposed, and Ptolemy accepted, that it was a being called a Prime Mover.

The Biography of Numbers

The Christians were just starting their religion in Ptolemy's time. They assigned the role of Prime Mover to their version of God. In their minds, then, and the minds of Christians who followed them, Aristotle and Ptolemy had done nothing less than prove God's existence.

From time to time Christian scholars pointed out holes in the theories. But Aristotle's and Ptolemy's universe nonetheless dominated Western (and later Muslim) thought for close to 1,500 years.

Zero, the tool capable of proving Ptolemy wrong, remained exiled from the ponderings of thinkers for the first millennium of that time. And when zero returned, powerful people gave it a hostile welcome.

Year Zero

Around the year 525, the monk Dionysius Exiguus, or Dennis the Small, created a table showing the dates of future Easter Sundays. To do so, Dionysius invented a new way of numbering years based on the life of Jesus Christ. The era before Christ's birth was thereafter considered B.C. The years after became A.D., for anno domini, "in the year of our Lord." (Today, the preferred term among scholars is BCE for B.C. and CE for A.D—BCE stands for Before the Common Era and CE for the Common Era.)

But the monk, like all Europeans in 525, remained ignorant of zero. In his system, 1 B.C. ended and the next day 1 A.D. began. No Year Zero separated the eras. Take the example of a woman born in 10 B.C. who died on her birthday in 10 A.D. Instinct tells us she was twenty at the time of death. Without that zero year, though, she actually died at age nineteen.

It took almost two hundred years, but Dionysius's system spread throughout Europe. It remains in use today, missing Year Zero and all.

CHAPTER 3

Long COUNTS and INDIAN NUMBERS

The Biography of Numbers

Though the Maya of Mesoamerica left behind monuments, pyramids, and other buildings, we know relatively little of them.

But archaeologists have uncovered a long record of remarkable scientific achievements. Mayan astronomers, for example, obsessively studied the movements of celestial bodies. Mayan writing makes it clear astronomers observed Venus with particular interest. Leaders may have planned their wars to coincide with Venus rising. The planet's appearance probably determined the best time to crown a new king.

Chichén Itzá Mayan observatory

ZERO

Calendars

Astronomy and mathematics merged in the Mayans' many elaborate calendars. Calculating time would lead the Maya to their independent discovery of zero.

Archaeologists have learned the Mayan script from inscriptions in stone. The script appears as carvings or is painted on religious structures and in ceremonial areas of cities, in places where Mayan priests lived, and in codices, folding books made of bark paper.

Much of the writing to survive exists on **stelae**, tall stone (often limestone) monuments. Mayans raised a stela to commemorate an event. Inscriptions on the monuments reveal the Maya used different calendars for different purposes. They divided their solar year calendar, called *tun*, into eighteen months of twenty days each. Five so-called nameless or unlucky days at the end made the calendar equal 365 days.

Stelae 5 at Takalik Abaj, El Asintal, Retalhuleu, Guatemala, showing an early example of a Long Count date

29

The Biography of Numbers

Mayan calendar

The *Tzolk'in*, a modern name for the Mayan religious calendar, tracked the days to observe sacred holidays and events. The planets Venus and Mars each had distinct calendars, too.

In addition, the Maya worked out a unique system, the Long Count, to compute dates far in the future. They started from their date of the world's creation. The calendar makers then counted forward. Mayan astronomers kept long counts of four hundred years (called a *baktun*), 8,000 years (*pictun*), and even longer, up to the 64 million-year-long *alautun*.

ZERO

The Maya wrote their dates in vertical rows. Because the Maya based their place value system in the number twenty, each row represented powers of twenty, in the way our system uses powers of ten. The lowest row indicated ones. The next up indicated twenties, or twenty times one. The third from the bottom indicated four hundreds, or twenty times twenty. And so on.

At some point, the calendar makers invented a placeholder to help clarify big numbers for readers. The Mayan version was a **glyph** shaped like a seashell, a hand gripping a seashell, or a distinct but more abstract shape.

Maya Zeros

A zero glyph clarified the numbers and therefore the dates. But the glyph shared the double slash's limits. Mayan mathematicians, like their Babylonian peers, never used the zero glyphs to calculate. Zero's leap into status as a true number would instead take place halfway around the world.

OC (logogram); named day 10 of the Tzolkin cycle

31

Brahmagupta

Ancient India had contact with Western civilizations going back to the Greeks and possibly Sumer. The influence of outside ideas on Indian mathematics remains under debate. But we know Indian ideas traveled in the opposite direction. Foremost among those ideas: the Indian system of nine distinct numbers, developed around or before 500 CE. These nine digits are the forerunners of the nine digits used by Americans and others today.

But the Indians also offered a tenth digit that changed the world.

The astronomer Brahmagupta was born around 598 CE. At about age thirty, he finished his first scientific masterpiece. The *Brahmasphutasiddhanta*, translated as the *Correctly Established Doctrine of Brahma*, was written in verse, the common method of written communication for an audience in Brahmagupta's time. Though focused on astronomy, the *Brahmaphutasiddhanta* covered many topics in mathematics. Today it is remembered as the earliest known text to discuss zero as a number.

To express zero in writing Brahmagupta placed a dot, called *bindu*, beneath another number. At times he called the dot *sunya*, or empty, an old word used to refer to an empty column on the abacus.

ZERO

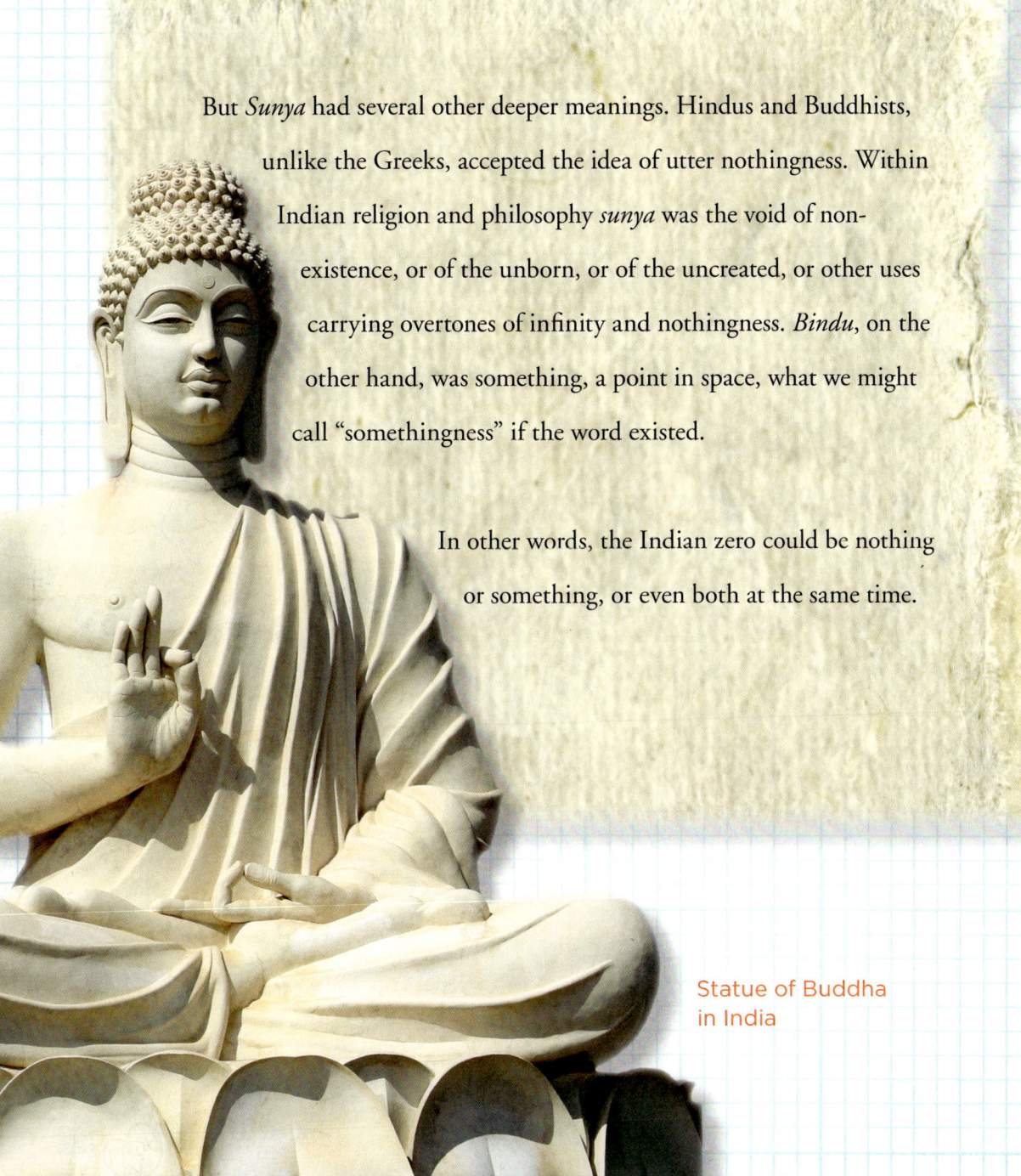

But *Sunya* had several other deeper meanings. Hindus and Buddhists, unlike the Greeks, accepted the idea of utter nothingness. Within Indian religion and philosophy *sunya* was the void of non-existence, or of the unborn, or of the uncreated, or other uses carrying overtones of infinity and nothingness. *Bindu*, on the other hand, was something, a point in space, what we might call "somethingness" if the word existed.

In other words, the Indian zero could be nothing or something, or even both at the same time.

Statue of Buddha in India

Using zero

Brahmagupta told readers how to add or subtract numbers to reach zero. In addition, he proposed the radical idea that adding zero to or subtracting zero from a number leaves the number unchanged. He went still further. Zero times zero equaled zero, he said, and a number times zero also equaled zero. The zero also gave Indians the tool needed to create gigantic numbers as well as miniscule numbers.

The idea of zero, as well as Brahmagupta's willingness to see numbers as **abstract** concepts, helped him explain another game-changing idea.

In earlier times, an equation like five minus eight had no answer. In fact, the Greeks would have considered it meaningless. But Brahmagupta believed in both an answer and a meaning to the equation. He laid out the concept of the negative number—he called it a debt—and the rules for working with negatives. His system accepted that an actual working number separated the negatives and positives. The number was zero.

ZERO

Arabian exchange

Muhammad, the prophet and founder of Islam, died in 632. Two years after his death, Arab armies under Islam's banner set out on a series of conquests that built an empire stretching from Spain to the frontiers of China.

Starting in 664 CE, Islamic armies raided into India on a regular basis. Contact with Indian civilization eventually introduced the Muslim world to zero.

Baghdad was founded in 762 as the new Islamic capital. The city immediately became the major learning center of the Muslim world. Muhammad ibn Musa al-Khwarizmi, born to a Persian family circa 780, joined the House of Wisdom, Baghdad's famed university-library-research center. Though also a geographer and astronomer, al-Khwarizmi is best remembered for his work in mathematics. His book on solving equations was a foundational text in the invention of algebra.

A statue of al-Khwarizmi in Khiva, Uzbekistan

The Biography of Numbers

al-Khwarizmi's *On the Calculations with Hindu Numerals* introduced the Indians' one through nine numerical system to the Islamic world. (Today we mistakenly refer to the digits as Arabic numerals.) At the same time al-Khwarizmi drafted the Indian zero into Muslim mathematics. *Sunya* became in Arabic *sifr*.

The new numbers, including zero, spread among the educated elite throughout the Islamic world. But it took until about the year 1000 for Muslim merchants and the rest of society to adopt the system.

European contact with Islamic peoples in both war and peace brought Muslim ideas to Europe. al-Khwarizmi's book on the misnamed Arabic numerals soon reached Islamic Spain. By the end of the 1100s English monks had copied it for themselves.

But the new system remained mostly obscure in Europe.

The Monk Eadwine
around 1150

Acquiring knowledge

Islam doggedly resisted the ideas of other faiths. But at the same time Muhammad's teachings encouraged Muslims to acquire knowledge in other fields. Muslims absorbed the cultural and intellectual traditions of the peoples they conquered. Knowledge from ancient civilizations in Egypt, Rome, and India found a place in Islamic universities and libraries.

CHAPTER 4

The MESSENGER

The Biography of Numbers

Leonardo Pisano Bigollo's father worked for the powerful Italian trading city of Pisa. His duties took him to Bugia, a Muslim port in North Africa. After joining his father, the young Leonardo attended a Bugia **accounting** school to learn business and Muslim mathematics.

"There," Leonardo wrote later, "when I had been introduced to the art of the Indians' nine symbols through remarkable teaching, knowledge of the art very soon pleased me above all else."

Leonardo spent his young adulthood traveling on business. But whenever possible he studied algebra and arithmetic with Arab mathematicians.

A statue of Leonardo Pisano Bigollo

ZERO

In the year 1200, the thirty-year-old Leonardo went home to Pisa. Two years later, he published a book on mathematics. *Liber Abaci* (Book of Calculation) introduced the Indian numerals to the intellectual elite of Europe. "The nine Indian figures are: 9 8 7 6 5 4 3 2 1," he stated. "With these nine figures, and with the sign 0, which the Arabs call *zephyr,* any number whatsoever is written . . ."

With that, he changed Western mathematics.

Map of Pisa

Leaning Tower of Pisa

The Biography of Numbers

Practical math

Leonardo, known today by the nickname Fibonacci, pitched his ideas to businessmen as well as scholars. Fibonacci crammed page after page of practical examples into *Liber Abaci* in order to demonstrate the everyday usefulness of the Indian system. Converting the various currencies of each city, for example, vexed Italian merchants, because every city minted its own money. Fibonacci took the reader through his word problems one step at a time, showing him or her where to place the numbers and how to use them.

Fibonacci's approach was an important part of convincing others to adopt Indian numbers. More broadly, he declared the superiority of Indian numbers to the inadequate Roman system (of X for 10, L for 50, and so on) that dominated in Europe but made calculations beyond addition and subtraction extremely difficult at best, and often impossible.

Roman numerals on a building

ZERO

Dome in Florence

Changing minds

Italy's city-states ran on business and finance. Once merchants and bankers understood Indian numerals—a process that took decades in some places, and centuries in others—Fibonacci's work revolutionized commerce. Suddenly men of business could easily calculate both the assets and debts on their balance sheets. The zero showed where one met the other.

Traditional institutions, especially governments, resisted the new system at first. The powerful city of Florence even outlawed Indian numerals, in part because it was so easy for criminals to change the new numbers on documents.

But money talked in thirteenth and fourteenth century Italy as assertively as it does today. Merchants and bankers continued to use Indian numbers. When doing so required them to evade the law, they sent coded messages called ciphers, a corruption of the Arabic *sifr*.

Battle for the Universe

At first, zero's usefulness gained it some acceptance. But the new numbers also raised a dangerous contradiction. Zero revived the idea of infinity, the enemy of Aristotle. Even before Fibonacci, astronomers had made observations that undermined the Aristotle-Ptolemy theory that all heavenly bodies revolved around the Earth. A rival idea, **heliocentric theory**, proposed the sun as the center of all things.

The Roman Catholic Church, still a dominant force in European intellectual life, defended Aristotle. The philosopher had, after all, proved the existence of the Christian God, or so the Church said.

Yet a number of monks and priests of independent mind openly declared their doubts about Aristotle, Earth's uniqueness, or the crystal spheres.

For a time they got away with it. By 1520, however, the **Protestant Reformation** had shaken Europe. With each year, more people joined Protestant Christian faiths breaking away from Catholicism. The Catholic Church, like many institutions under attack, suddenly considered tolerance of new ideas dangerous. Church leaders condemned any criticism of Aristotle's ideas.

ZERO

In time, zero won out. The Church may have banned the work of Nicholas Copernicus, the Polish astronomer who proved the Earth orbited the sun. But Copernicus's heirs investigated his ideas further. Their refinements to his theories made his ideas irrefutable.

Nicholas Copernicus

The process repeated on other important scientific questions. By the end of the 1600s, the revolution in mathematics brought about in part by zero had smashed Aristotle's and Ptolemy's crystal spheres into a thousand pieces.

The heliocentric theory according to Copernicus

45

Dangerous science

The Church campaign against heresy took an increasingly anti-science tone as the new Protestant religions challenged Church supremacy.

Scientists determined to study forbidden subjects faced real threats like loss of work and imprisonment—and worse. In the year 1600, the Church burned Italian friar Giordano Bruno at the stake for his belief in the infinity of the universe and his rejection of Aristotle. Galileo Galilee, a scientist of similar mind, agreed to keep quiet after Church representatives "suggested"

Giordano Bruno statue

he cease his investigations and showed him a torture chamber to punctuate their point.

The Church even claimed the power to send so-called heretics to hell.

Whereas Italian cities banned Arabic/Indian numerals for practical reasons, some members of the Church declared the numbers to be the work of Satan. Nothing else seemed to explain how math—once a grueling chore done by a small number of specialists working with Roman numerals—had been transformed into a handy tool that even a layman could figure out with a bit of training.

CHAPTER 5

The ZERO Revolutions

The Biography of Numbers

The acceptance of zero brought about changes in science and technology that we take for granted. Indeed, zero made today's science and technology possible.

Rene Descartes, a French philosopher-mathematician, lived and worked in the Netherlands, a more tolerant country than Galileo's Italy, and one where the Church had less influence. Descartes used zero in his groundbreaking Cartesian coordinate system. The system's origin point of (0,0) lay where the vertical axis met the horizontal axis.

A bust of Rene Descartes

ZERO

Calculus

Within a few years of entering college, Isaac Newton (1642-1726) had made advances in mathematics that laid the groundwork for calculus, the language of science and engineering and, as far as human beings can speak it, of the universe itself. Zero made calculus possible.

Newton's towering 1687 work *Principia,* or *The Mathematical Principles of Natural Philosophy,* showed the author using calculus as a tool to investigate the motions of planets and many other questions.

A print of Isaac Newton created by William Blake

German intellectual and diplomat Gottfried Wilhelm Leibniz initially studied law and later worked in several fields. But Leibniz also trained in physics and mathematics, often on his own.

During his period of self-education he corresponded with Newton. By 1676, and perhaps earlier, Leibniz also invented calculus. Newton accused Leibniz of stealing his work, work that at the time remained unpublished. Leibniz countered with the same charges. Their methods differed in many ways, however, and most historians grant Leibniz the honor of co-discovering calculus on his own.

The Biography of Numbers

Absolutes

Since the 1700s, scientists have worked with zero to ponder nature's extremes. William Thomson, also known as Lord Kelvin, experimented with shrinking a container of gas to a point where it took up the smallest possible space. To do so meant removing all of the energy of the gas (or any object being subjected to the same process). At that point, the atoms involved ceased moving. Kelvin declared 273 degrees zero the coldest possible temperature. He called it absolute zero.

Physicists in the twentieth century discovered an absolute zero of another kind: the total nothingness of the black hole.

Stars are balls of heated gas. Their massive size should make them prone to collapse under gravity. But the energy released by the hydrogen at the star's center counterbalances the gravitational forces.

Lord Kelvin

ZERO

A black hole simulated

In time, however, the hydrogen fuel runs low. As it does, the star collapses in on itself.

With certain giant stars, gravity condenses the star's mass to such an extreme degree that its very atoms fall apart. Finally, the star's matter then collapses into . . . well, a sort of zero in space called a black hole centered around a core called a singularity.

A black hole *exists*. It affects space and time. It possesses mass. Its irresistible gravitational pull draws everything toward it; and even light, once beyond a boundary called the event horizon, cannot escape. Yet at the same time the black hole *does not seem to exist* because it takes up no space, not even the *bindu* of the ancient Indians.

In a sense most of us think of the idea of black hole in a way that shares one thing with zero. It's possible to perceive both black holes and zero to be nothing. At the same time, that belief could not be further from the truth.

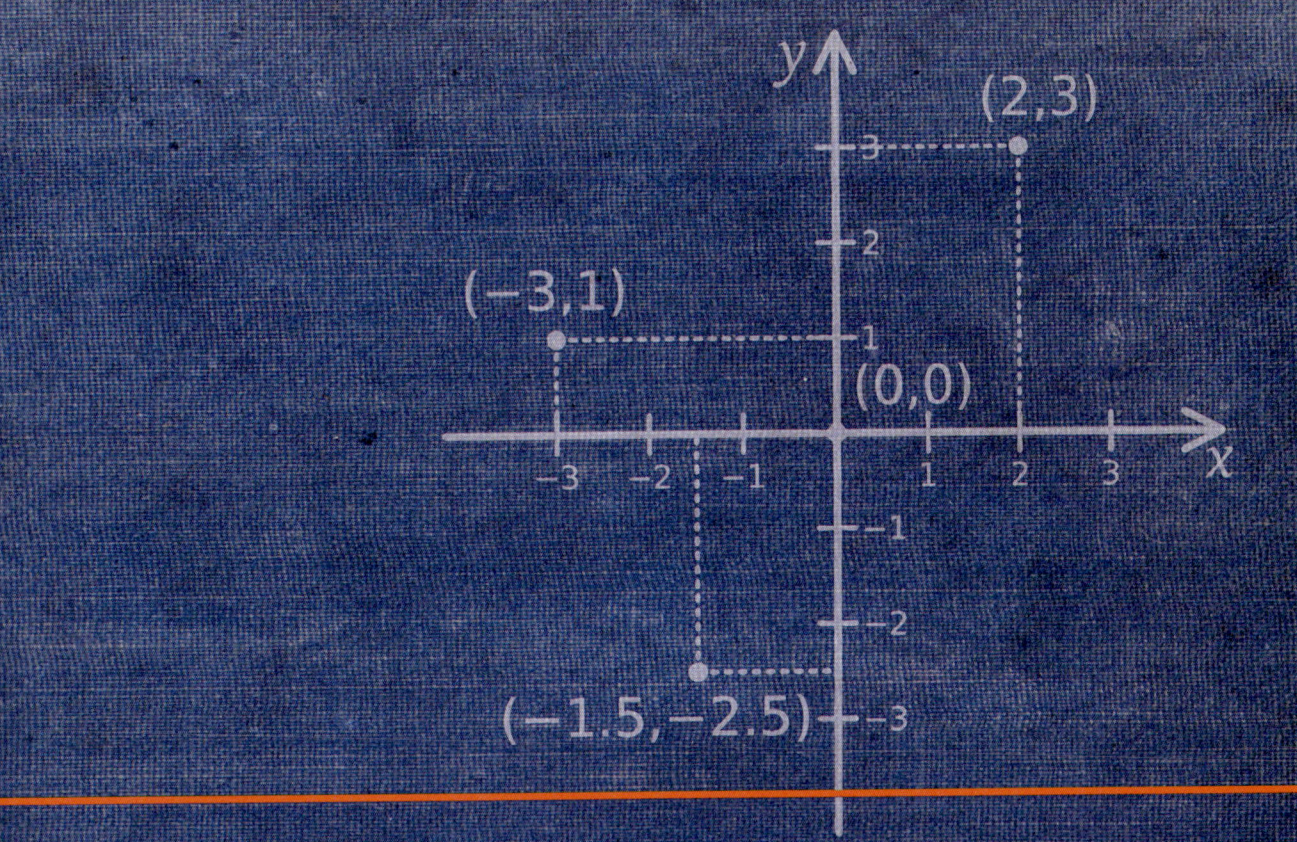

Cartesian coordinate system

The Cartesian system gave mathematicians a straightforward way of describing geometric shapes using equations. The Cartesian coordinate system remains at the heart of any discipline, from engineering to computer graphics, based in geometry. Ironically, Descartes made his breakthroughs while maintaining a belief in Aristotle. He simply reinterpreted the old philosopher's system using the scientific discoveries of his own time.

Timeline

c. 1500 BCE	*Nfr* placeholder possibly used in Egypt
c. 500 BCE	First use of the double wedge, a placeholder zero, in Babylonian documents
c. 384 BCE	Aristotle born in Stageira, Greece
100s CE	Ptolemy discusses the structure of the universe in his *Almagest*
c. 250	Maya enter Classic Period of peak scientific achievement
c. 500	Indians develop a number system with nine digits
c. 525	Dionysius Exiguus creates the B.C.-A.D. dating system

628	Brahmagupta elaborates on zero in the *Brahmasphutasiddhanta*
c. 825	al-Khwarizmi's *On the Calculation with Hindu Numerals* brings zero to the Muslim world
1202	Fibonacci publishes *Liber Abaci* in Italy; zero enters European thought
1684	Gottfried Leibniz publishes his first paper on calculus
1687	Isaac Newton finishes *The Mathematical Principles of Natural Philosophy*
1973	Astronomer John Wheeler first uses the term "black hole"

Biographical sketches

Aristotle (c. 384 BCE-c. 322 CE)
http://space.about.com/od/astronomerbiographies/a/aristotlebio.htm

One of the giants of Western intellectual history, Aristotle wrote on a vast range of topics in science and philosophy. Many of his works have been lost. His model of the universe influenced Ptolemy and the Catholic Church defended it during medieval and Renaissance times. Later in life Aristotle tutored the future Alexander the Great and founded a school outside Athens.

Claudius Ptolemy (c. 90-168)
http://www.universetoday.com/81048/ptolemy-astronomy/

Considered one of the greatest scientists of the Roman Age, Ptolemy published the *Almagest* in the mid-100s. In it he explained the movement of heavenly bodies in mathematical terms and detailed a model of an Earth-centered universe composed of concentric crystal spheres. The Ptolemaic system, borrowed in part from Aristotle, dominated European and Muslim astronomy for almost 1,500 years.

Dionysius Exiguus (c. 470-c.544)
http://www.encyclopedia.com/topic/Dionysius_Exiguus.aspx

Dionysius traveled from his monastery in Eastern Europe to Rome just before 500 and became a respected scholar. In 525, the pope requested he create tables to find dates for future Easter Sundays, and in related work Dionysius invented the Christian B.C.-A.D. calendar still in wide use today, though, the preferred terms are now BCE and CE.

Brahmagupta (598-c. 665)
http://www.britannica.com/EBchecked/topic/77073/Brahmagupta

Born in northwest India, Brahmagupta joined an astronomical observatory in his thirties and wrote influential texts on both astronomy and mathematics. In addition to formulating ideas for zero, Brahmagupta made advances in topics like equations and algebra, and offered rules for the calculation and use of negative numbers.

Mohammed ibn-Musa al-Khwarizmi (c. 780-c. 850)

http://www.britannica.com/EBchecked/topic/317171/al-Khwarizmi

al-Khwarizmi lived and worked in Baghdad. One of the founders of algebra, al-Khwarizmi wrote the book introducing the Indian concept of zero, and the number itself, to the Muslim world. In his other work al-Khwarizmi drew on ancient sources of many cultures to redraw maps of the Earth and create tables useful for astronomical observation.

Fibonacci, or Leonardo Pisano Bigollo (c. 1170-c. 1250)

http://www.britannica.com/EBchecked/topic/336467/Leonardo-Pisano

The son of a merchant and government official, Leonardo learned "Indian numerals" studying mathematics in Islamic schools in North Africa. His *Liber Abaci* introduced the number system to a wide European audience. In time he corresponded with the Holy Roman Emperor and esteemed scholars in the emperor's courts. A statue of him stands in Pisa today.

Isaac Newton (1642-1726)

http://www.bbc.co.uk/history/historic_figures/newton_isaac.shtml

Newton survived a rough upbringing, depression, and prickly relationships with peers to become the foremost scientist of his time. Best known for his work on gravity and light, Newton also made mathematical breakthroughs that led to calculus. But his habitual reluctance to publish his work kept many of his ideas, including those on calculus, from the public for years or decades.

Gottfried Leibniz (1646-1716)

http://scienceworld.wolfram.com/biography/Leibniz.html

Something of a child prodigy, Leibniz spoke many languages and practiced in disciplines ranging from law to biology to philosophy. His groundbreaking work on calculus led to a feud with Isaac Newton. His work on mathematical notation, meanwhile, remains in use. Leibniz also designed the Leibniz wheel, an essential part of the non-electronic calculators used into the twentieth century, and he contributed to the binary number system at the core of modern computer science.

Glossary

abacus (A-be-kes)
A calculating tool one uses by sliding counters like stones or beads on wires set within a frame.

abstract (ab-STRAKT)
A theory that (often) exists outside of any practical application.

accounting (eh-KAUN-ting)
The system of tracking financial transactions.

epicycle (E-pa-SY-kel)
The orbit of a planet around a center that itself moves around the edge of a larger circle.

glyph (GLIF)
A symbolic mark or character.

heliocentric theory (HE-lee-o-SEN-trik THIR-ee)
The theory that the Earth revolves around the Sun.

heresy (HER-eh-see)
An opinion or theory that goes against the stated beliefs of one's religion.

hieroglyph (HI-ro-glif)
A picture or shape that stands for a word in a hieroglyphic writing system.

infinity (in-FI-ne-tee)
A number of amount without limit or end.

Islam (IS-lam)
The faith of Muslims based on the teachings of the prophet Muhammad.

place value (PLAYSE VALL-yu)
A system that assigns a value to a single digit based on its position in a number. In the Western place value system based on tens, the digit furthest right in a whole number represents ones. The digit to its left represents tens (10 times 1). The digit to its left represents hundreds (10 times 10). Other cultures invented system based in five, twenty, or even sixty.

Protestant Reformation (PRA-tes-tent RE-fer-MAY-shun)
A movement begun in 1517 by clergymen, nobles and other who wished to reform, but eventually broke away from, the Catholic Church.

Ptolemaic theory (TA-le-MAY-ik THIR-ee)
Ptolemy's theory that the Earth is at the center of universe and the other heavenly bodies revolve around it.

scribe (SKRIBE)
A job occupation practiced by those who specialized in writing out books, documents, and other records by hand.

stelae (STEE-lay)
Carved or inscribed stone slabs used to commemorate an event or events.

sunya (SUNE-ya)
A Sanskrit word originating in India that means "empty" but also the Indian numeral zero, as well as various states of being in the Hindu and Buddhist belief system.

vacuum (VA-kyum)
A space that is devoid of matter.

Bibliography

Books

Barrow, John D. *The Book of Nothing: Vacuums, Voids, and the Latest Ideas About the Origins of the Universe.* New York: Vintage, 2002.

Crumpacker, Bunny. *Perfect Figures: The Lore of Numbers and How We Learned to Count.* New York: St. Martin's, 2007.

Demarest, Arthur, *The Ancient Maya: The Rise and Fall of a Rainforest Civilization.* New York: Cambridge University Press, 2004.

Devlin, Keith. *The Man of Numbers: Fibonacci's Arithmetic Revolution.* New York: Walker and Company, 2011.

Fibonacci, and L. E. Sigler, trans. *Fibonacci's Liber Abaci: Leonardo Pisano's Book of Calculation.* New York: Springer-Verlag, 2003.

Ifrah, George, and David Bello, trans. *The Universal History of Numbers.* New York: Wily, 2000.

Kaplan, Robert. *The Nothing That Is: A Natural History of Zero.* New York: Oxford University Press, 2000.

McLeish, John. *Number.* New York: Fawcett Columbine, 1991.

Morley, Sylvanus Griswold, and George W. Brainerd. *The Ancient Maya, third edition.* Stanford, CA: Stanford University Press, 1968.

Posamentier, Alfred S., and Ingmar Lehmann. *The (Fabulous) Fibonacci Numbers.* New York: Prometheus, 2007.

Rudman, Peter S. *How Mathematics Happened: The First 50,000 Years.* New York: Prometheus, 2007.

Seife, Charles. *Alpha and Omega: The Search for the Beginning and End of the Universe.* New York: Penguin, 2003.

———. *Zero: The Biography of a Dangerous Idea.* New York: Viking, 2000.

Sharer, Robert J., and Loa P. Traxler. *The Ancient Maya, sixth edition.* Stanford, CA: Stanford University Press, 2006.

Smith, D. E. *History of Mathematics, Volume 1.* New York: Dover, 1958.

Steel, Duncan. *Marking Time: The Epic Quest for the Perfect Calendar.*

Stewart, Ian. *Taming the Infinite: The Story of Mathematics from the First Numbers to Chaos Theory.* London: Quercus, 2008.

Temple, Robert. *The Genius of China.* Rochester, VT: Inner Traditions, 2007.

Periodicals and online

Arsham, Hossein. "Zero in four dimensions: Cultural, historical, mathematical, and psychological perspectives." Personal Web site. http://home.ubalt.edu/ntsbarsh/zero/ZERO.HTM#rintro.

Bibliography continued

Blume, Anna. "Maya concepts of zero." *Proceedings of the American Philosophical Society* 155, no. 1:51-88.

Casselman, Bill. "All for nought." *American Mathematical Society Feature Story*. http://www.ams.org/samplings/feature-column/fc-current.cgi.

Gascoigne, Bamber. "History of the Abacus." HistoryWorld. From 2001, ongoing. http://www.historyworld.net/wrldhis/PlainTextHistories.asp?historyid=095.

Hirst, K. Kris. "Maya writing got early start." *Science Online,* January 6, 2006. http://news.sciencemag.org/sciencenow/2006/01/06-02.html.

Joseph, George G. "A brief history of zero." *Iranian Journal for the History of Science,* 6 (2008): 37-48.

Law, Steven. "A brief history of numbers and counting, part two: Indian invention of zero was huge in development of math." *Deseret News,* August 6, 2012. http://www.deseretnews.com/article/865560133/A-brief-history-of-numbers-and-counting-Part-2-Indian-invention-of-zero-was-huge-in-development-of.html?pg=all.

Matson, John. "The origin of zero." *Scientific American,* August 21, 2009. http://www.scientificamerican.com/article.cfm?id=history-of-zero.

Schmandt-Besserat, Denise, and Wayne M. Senner, eds. "Two precursors to writing: Plain and complex tokens." From *The Origin of Writing*. Lincoln, NE: University of Nebraska Press, 1991. http://en.finaly.org/index.php/Two_precursors_of_writing:_plain_and_complex_tokens.

Wallin, Nils-Bertil. "How was zero discovered?" *YaleGlobal,* November 19, 2002. http://yaleglobal.yale.edu/about/zero.jsp.

Web sites

http://www.bbc.co.uk/programmes/p004y254 [1]

British Broadcasting Corporation. "Zero." _In Our Time._ May 13, 2004. A group of experts discuss the history of zero on a 2004 radio show first broadcast in Great Britain. The site also offers shows on Indian mathematics and other math-related topics.

http://www.youtube.com/watch?v=gulApUKih2w

British Broadcasting Corporation. *The Story of Numbers*. Undated. Via YouTube. A humorous history of zero (and one) hosted by British writer-comedian Terry Jones.

http://www.bbc.co.uk/programmes/p004y254

British Broadcasting Corporation Radio, "Zero." *In Our Time,* May 13, 2004. A discussion of zero's origins and how it changed the world by a group of authors and mathematicians.

http://deyoung.famsf.org/files/collectionicons/index1.html

De Young Fine Arts Museums of San Francisco. An online gallery that offers high definition photos of a Maya stela as well as information on a wide variety of Mayan artworks.

http://www.npr.org/2011/07/16/137845241/fibonaccis-numbers-the-man-behind-the-math

National Public Radio. "Fibonacci: The Man Behind the Numbers." *Weekend Edition.* July 16, 2011.

http://video.pbs.org/video/2036276385/

Public Broadcasting System. "Galileo's Battle for the Heavens." *Nova,* October 28, 2002. The story of how scientists like Galileo Galilee and others undermined and ultimately overthrew Aristotle's view of the universe.

Index

abacus, 12, *12*
absolute zero, 52
al-Khwarizmi, Muhammad ibn Musa, 35–36, *35,* 59, *59*
Arabic numbers, 36, 47
Aristotle, 20, *20,* 22, 24, 44–46, 58, *58*
Bigollo, Leonardo (Fibonacci), 40–44, *40,* 59, *59*
black hole, 52–54, *53*
Brahmagupta, 32–34, 58, *58*
Bruno, Giordano, 46, *46*
calendars, 25, 29–31, *30*
Cartesian coordinates, 50, 55, *55*
Copernicus, Nicholas, 45, *45*
cuneiform, 11, *11,* 13, 15
Descartes, Renee, 50, *50,* 55
Dionysius Exiguus (Dennis the Small), 25, 58, *58*
double wedge, 14
empty space, 14, 32
Galilee, Galileo, 46–47
glyph, 31, *31*
heliocentric theory, 44, *45*
India, numbering system of, 32, 40–43, 47
infinity, concept of, 18, 23, 44, 46

Leibniz, Gottfried, 51, 59, *59*
negative numbers, 34
Newton, Isaac, 51, *51,* 59, *59*
nfr, 10–11
omicron, 18
place holder, 18
place value, 14, 31
Ptolemy, Claudius, 20–24, *20,* 45, 58, *58*
Roman numbers, 42, *42,* 47
science and the church, 24, 28, 44–47, 50
sifr, 36, 43
sunya, 32–33, 36
Thomson, William (Lord Kelvin), 52, *52*
zephyr, 41
zero
    acceptance of, 50
    in daily life, 10, 36, 43–44
    early concepts of, 9, 14
    as a number, 31–32
    symbols for, 10–11, 18, 31–32, 36, 41
    use of in science and math, 34–35, 42–43, 45, 50–54

Photo credits

All images used in this book that are not in the public domain are credited in the listing that follows:

8:	Courtesy of Evans-Amos
9:	prehistoric sketch: Courtesy of
10:	Courtesy of Ricardo Liberato
11:	Courtesy of Matt Neale of UK
13:	Courtesy of Aziz1005
22-23:	Courtesy of Natiional Aeronautics and Space Administration
28:	Courtesy of Bruno Girin
29:	Courtesy of Simon Burchell
30:	Courtesy of theilr
32-33:	Courtesy of Purshi
35:	Courtesy of Euyasik
40:	Courtesy of Hans-Peter Postel
41:	Leaning Tower of Pisa: Courtesy of Aaron Logan
42:	Courtesy of Wolfgang Sauber